# Contemporary Mathematics

## A Tutorial for Recitation Sessions

### Third Edition

**Joseph L. Yucas**
**Walter D. Wallis**
*Department of Mathematics*
*Southern Illinois University*

Cover image courtesy of Corel

**Kendall Hunt**
publishing company

Copyright © 1999, 2006, 2010 by Kendall/Hunt Publishing Company

ISBN 978-0-7575-7228-9

All rights reserved. No part of this publication may be reproduced,
stored in a retrieval system, or transmitted, in any form or by any means,
electronic, mechanical, photocopying, recording, or otherwise,
without the prior written permission of the copyright owner.

Printed in the United States of America
10  9  8  7  6  5  4  3  2  1

# CONTENTS

### 1. GRAPHS AND EULER CIRCUITS
Worksheet 1 .................................. 1
Sample Exam 1 ............................. 9

### 2. HAMILTONIAN CIRCUITS AND THE COUNTING PRINCIPLE
Worksheet 2 .................................. 21
Sample Exam 2 ............................. 29

### 3. TREES AND GRAPH COLORING
Worksheet 3 .................................. 39
Sample Exam 3 ............................. 45

### 4. SCHEDULING AND BIN PACKING
Worksheet 4 .................................. 55
Sample Exam 4 ............................. 63

### 5. DESCRIPTIVE STATISTICS
Worksheet 5 .................................. 73
Sample Exam 5 ............................. 81

### 6. PROBABILITY, NORMAL CURVES
Worksheet 6 .................................. 91
Sample Exam 6 ............................. 97

### 7. SAMPLING DISTRIBUTIONS, CONFIDENCE INTERVALS, CODE NUMBERS
Worksheet 7 .................................. 107
Sample Exam 7 ............................. 113

## 8. Error-Correcting Codes

Worksheet 8 .................................... 123
Sample Exam 8 ................................ 129

## 9. Voting and Geometric Growth

Worksheet 9 .................................... 137
Sample Exam 9 ................................ 143

## 10. Review

Final Worksheet ................................ 153
Sample Final Exam ............................ 167

## 11. Answers to the Sample Exams    195

## Some Lecture Notes    199

# GRAPHS AND EULER CIRCUITS

## WORKSHEET 1

1. Draw a graph that has four vertices with valence three and one vertex with valence two.

2. Find an Euler circuit on the graph given below.

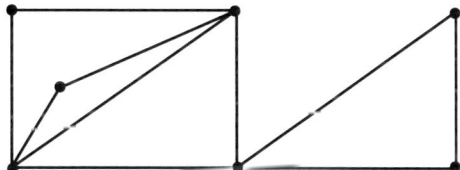

3. Does the graph given below have an Euler circuit?

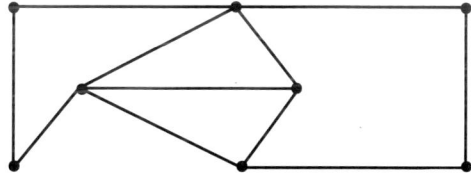

1

4. Consider the street network given below. Draw a graph that could be used for finding an efficient route for a paper delivery boy.

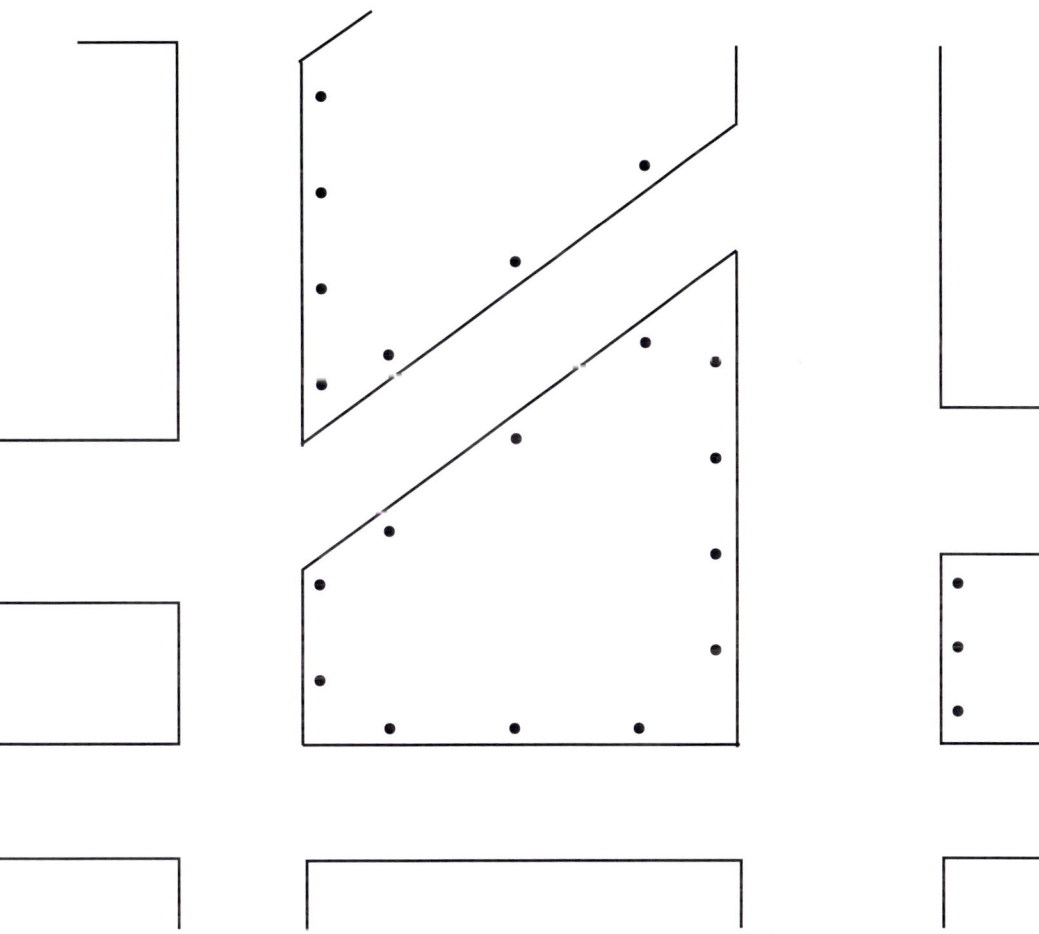

5. Is the graph given below a connected graph?

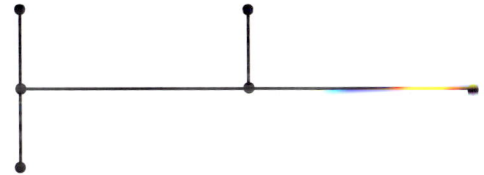

6. Eulerize the graph given below.

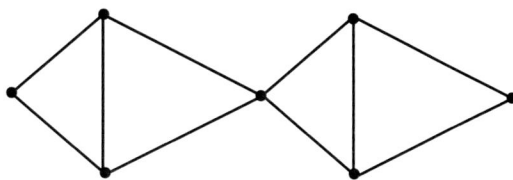

7. Is every graph with five vertices and four edges connected?

8. Would an Euler circuit be useful for a salesman who must visit six cities and then return home?

9. Eulerize the graph given below.

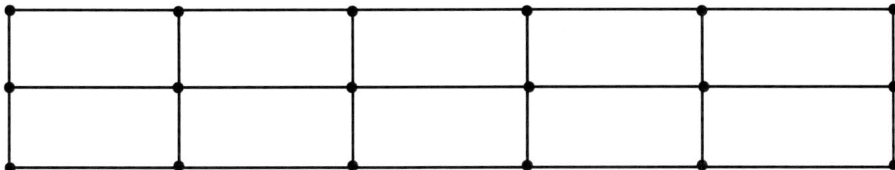

10. Draw a graph that has a circuit but which has no Euler circuit.

# SAMPLE EXAM 1

1. Every graph that has an Euler circuit must have an even number of vertices.

    A) True

    B) False ✓

2. What is the valence of vertex E in the following graph?

    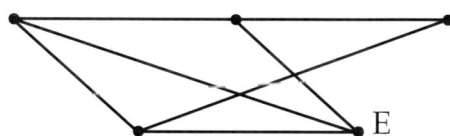

    A) 3

    B) 4

    C) 5

    D) 6

3. Is the following graph a connected graph?

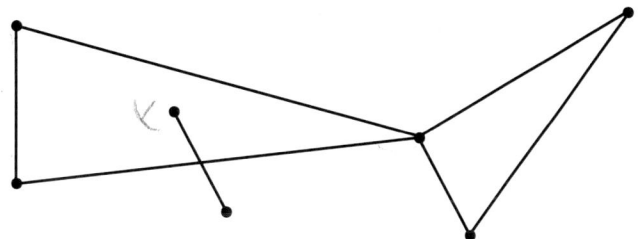

A) Yes

B) No ✓

4. A graph that has ten vertices and five edges cannot be a connected graph.

A) True ✓

B) False

5. Which of the graphs below have Euler circuits?

   1. 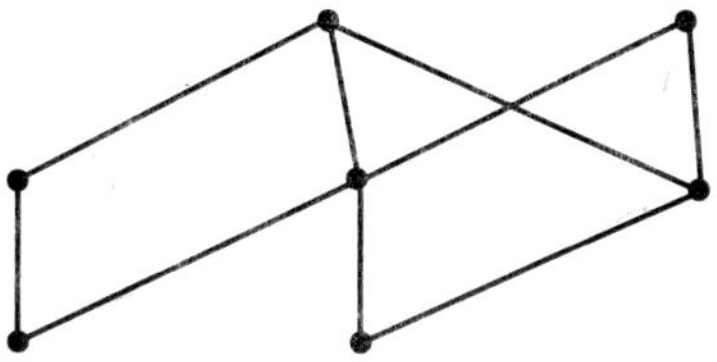
   
   2.

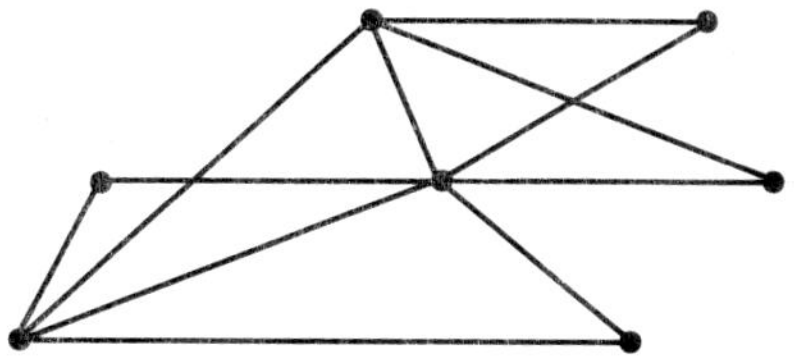

   A) 1 only
   
   B) 2 only ✓
   
   C) Both
   
   D) Neither

6. The path given by the numbered sequence of edges on the graph below is not an Euler circuit. Why?

   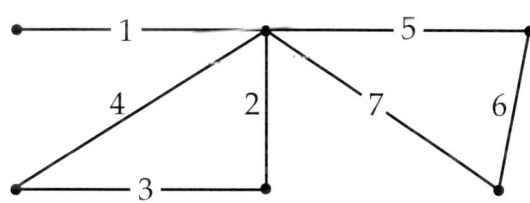

   A) It hits a vertex more than once.
   
   B) It does not cover every edge.
   
   C) It is not a circuit.   Starts & ends at the same vertex
   
   D) None of the above.

7. The graph below does not have an Euler circuit. Which edge could be duplicated so that the new graph will have an Euler circuit?

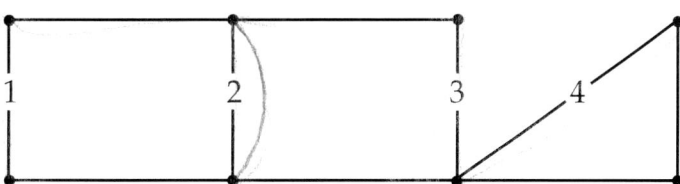

A) 1

B) 2

C) 3

D) 4

8. Would an Euler circuit be useful for a painting crew whose job is to paint the center lines of each street in a city?

A) Yes

B) No

9. To Eulerize the graph given below, what's the fewest number of edges that would have to be duplicated?

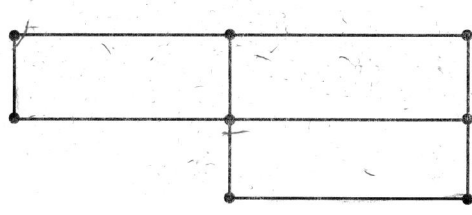

A) 0

B) 1

C) 2

D) 3

10. Suppose a graph has 10 vertices of odd valence. What is the absolute fewest number of edges that could be duplicated to Eulerize the graph?

A) 1

B) 5

C) 3

D) 10

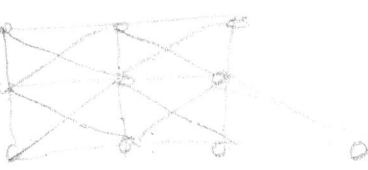

11. Consider the street network given below. How many vertices would the graph that models this network have?

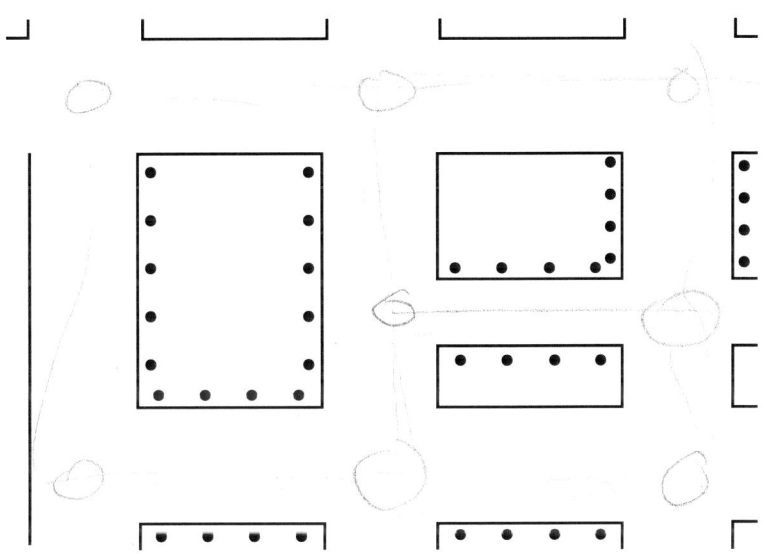

A) 8

B) 9

C) 10

D) 11

12. How many edges would the graph of Problem 11 have?

A) 8

B) 9

C) 10

D) 11

# HAMILTONIAN CIRCUITS AND THE COUNTING PRINCIPLE

## WORKSHEET 2

1. How many 3-digit numbers are there that start with an odd digit and end with an odd digit?

2. How many Hamiltonian circuits are there in the complete graph on ten vertices?

3. For the graph given below, use the nearest-neighbor algorithm starting at A to find a route for a traveling salesman.

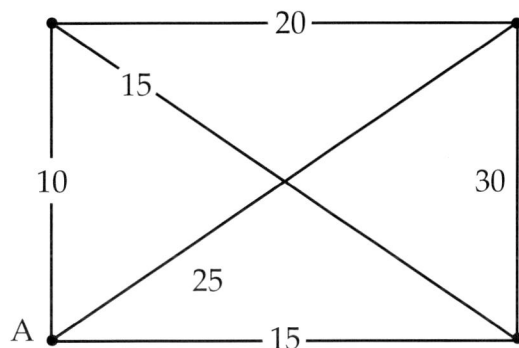

4. For the graph given in Problem 3, use the brute-force algorithm to find the best route for a traveling salesman.

5. For the graph given in Problem 3, use the sorted-edges algorithm to find a route for a traveling salesman.

23

6. How many distances would have to be measured to create a mileage grid showing the distances between 25 cities?

7. Use the sorted-edges algorithm to find the cheapest Hamiltonian circuit for the following graph.

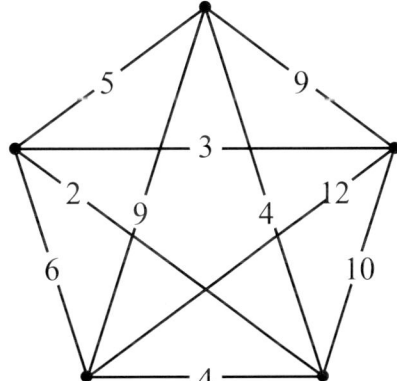

8. Use the nearest-neighbor algorithm on the graph given in Problem 7 starting at the top vertex.

9. Construct a graph that has no Hamiltonian circuit but which has an Euler circuit.

10. Your security code consists of a letter (from A to Z) followed by a 2-digit number (from 00 to 99). How many codes are possible?

# SAMPLE EXAM 2

1. Which of the following paths is a Hamiltonian circuit on the graph given below?

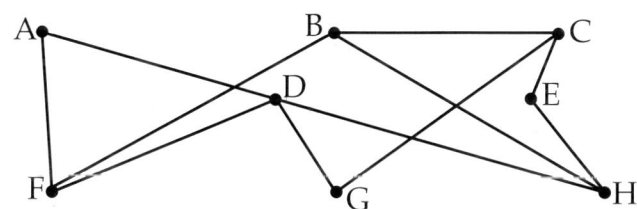

A) ADGCBEHFA

B) ADHECGDFA

C) AFBHECGDA

D) ADFBCEHDA

2. In applying the brute-force algorithm to the traveling salesman problem on seven cities, how many tours would have to be checked?

A) 360

B) 1224

C) 2520

D) 2864

29

3. Find the cost of the Hamiltonian circuit obtained by applying the sorted-edges algorithm to the graph given below.

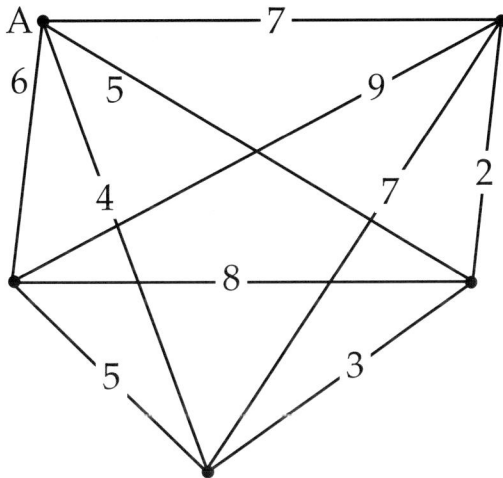

A) 25          C) 27

B) 24          D) 28

4. Find the cost of the Hamiltonian circuit obtained by applying the nearest-neighbor algorithm starting at vertex A to the graph given in Problem 3.

A) 25          C) 27

B) 24          D) 28

5. In a map, the roads from Town A to other towns are of lengths 13, 15 and 18 miles. Using the shortest-edges algorithm starting at town A, which road will be traveled first?

   A) The 13 mile road

   B) The 15 mile road

   C) The 18 mile road

   D) There is not enough information

6. The code for your bicycle lock consists of three non-zero single-digit numbers (that is, three choices from 1, 2, 3, 4, 5, 6, 7, 8, 9). You may not repeat a digit. How many possible codes are there?

   A) 24       B) 27       C) 504       D) 729

7. For which of the following would it be helpful to use Traveling Salesman algorithm?

   A) A city wants to plan its snow removal routes.

   B) A company needs to connect all its computers to one server.

   C) An airline has to monitor all its check-in procedures.

   D) None of these

8. A cafe offers four entrees, five side dishes, and three desserts. A complete meal consists of one of each. How many different complete meals are there?

    A) 12

    B) 54

    C) 48

    D) 60

9. How many possible different answer sheets are there for a five question true-false test?

    A) 16

    B) 32

    C) 26

    D) 48

10. The nearest-neighbor algorithm gives the same answer, no matter which starting vertex is used.

    A) True

    B) False

11. Suppose a graph has 8 vertices, and each vertex has valence 4. If the graph has a Hamiltonian circuit, how many edges will the circuit contain?

   A) 7                B) 8                C) 12                D) 16

12. Which of the following statements concerning algorithms for the Traveling Salesman Problem is accurate?

   A) The Sorted Edges algorithm always gives the optimum answer but the Nearest Neighbor algorithm does not.

   B) The Nearest Neighbor algorithm always gives the optimum answer but the Sorted Edges algorithm does not.

   C) Both these algorithms always give the optimum answer.

   D) Each algorithm sometimes fails to give the optimum answer.

# TREES AND GRAPH COLORING

## WORKSHEET 3

1. In each case, why is the graph not a tree?

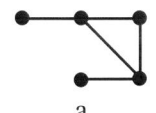

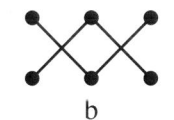

    a   b   c

2. Draw a tree with five vertices, three of which have valence 1.

3. A connected graph has six vertices. How many edges are there in a spanning tree for this graph?

4. Construct a spanning tree on the graph given below.

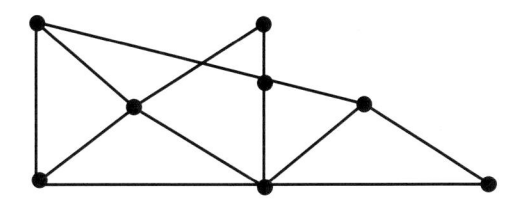

5. A graph consists of four vertices, each joined to each other. How many colors are needed for a proper coloring?

6. The diagram shows part of a graph in which Kruskal's algorithm is being applied. The heavy edges have been chosen. Which edge will be chosen next?

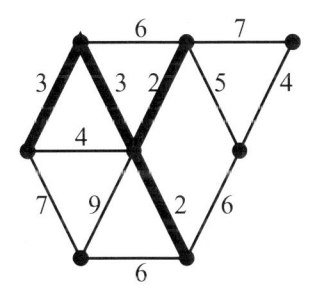

41

7. Consider the graph given below. Find a minimal-cost spanning tree using Kruskal's algorithm.

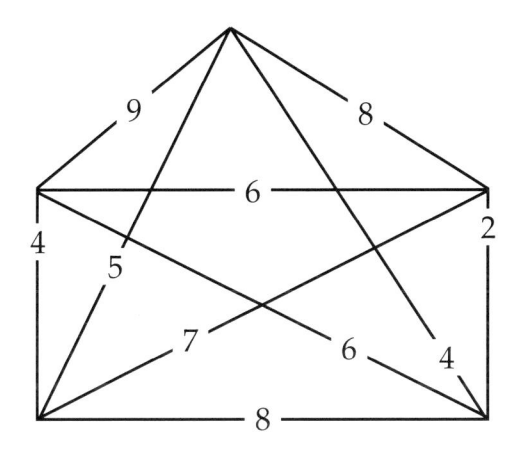

8. What is the chromatic number of the following graph?

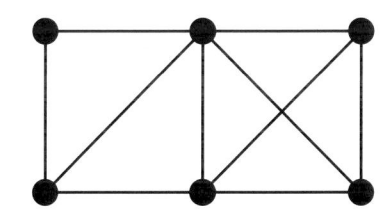

9. Suggest a practical situation for which it would be useful to find a spanning tree, but a Hamiltonian circuit would be of no use.

10. Suggest a practical situation for which it would be useful to find a vertex coloring, but a spanning tree would be of no use.

# Sample Exam 3

1. Which of the following is true?

    A) Every graph has a unique spanning tree.

    B) Every connected graph has a unique spanning tree.

    C) Every connected graph has a unique minimum-cost spanning tree.

    D) None of the above.

2. Suppose a graph has 8 vertices, and each vertex has valence 4. Assuming the graph has a spanning tree, how many edges will the tree contain?

    A) 7	B) 8	C) 12	D) 16

3. What is the cost of a minimal spanning tree in the following graph?

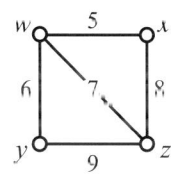

    A) 5	B) 18	C) 24	D) 35

4. Is the graph given below a tree?

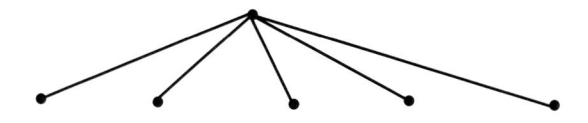

   A) Yes

   B) No

5. Suppose a tourist wishes to visit every museum in a particular city. Which of the following would be useful for him?

   A) A minimal-cost Euler circuit.

   B) A minimal-cost Hamiltonian circuit.

   C) A minimal-cost spanning tree.

   D) None of these.

6. Which of the following are proper colorings?

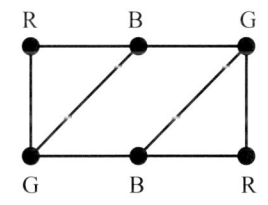

   A) I only

   B) II only

   C) Neither I nor II

   D) Both I and II

7. What is the chromatic number of this graph?

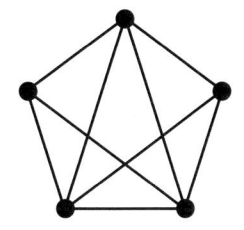

A) 2  C) 4
B) 3  D) 5

8. Apply Kruskal's algorithm to find a minimum-cost spanning tree in the following graph. What is the cost?

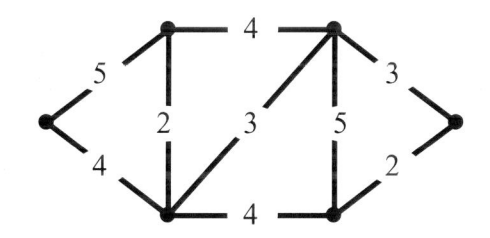

A) 10  C) 15
B) 14  D) 20

9. The table below shows chemical compounds that cannot be mixed without causing dangerous reactions. Which graph would be used to facilitate scheduling disposal of containers for the compounds?

|   | A | B | C | D | E |
|---|---|---|---|---|---|
| A |   | X |   |   | X |
| B | X |   | X |   | X |
| C |   | X |   | X |   |
| D |   |   | X |   |   |
| E | X | X |   |   |   |

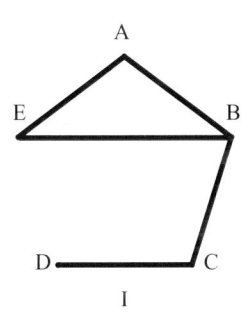

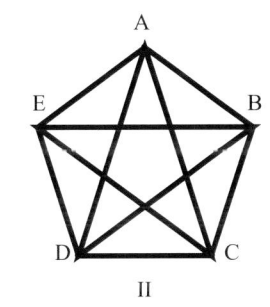

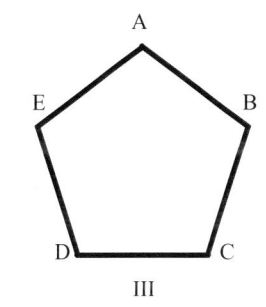

I   II   III

A) I   B) II   C) III   D) None of them

10. The city needs to inspect the roads after a storm, to remove tree branches. Which method will be most helpful to them?

A) Construct an Euler circuit or minimal Eulerization.

B) Construct a minimum-cost Hamiltonian circuit.

C) Construct a minimum-cost spanning tree.

D) Construct a vertex coloring in the fewest possible colors.

11. A pet store owner wishes to keep several varieties of tropical fish in a small number of tanks. He must not put two species in the same tank. Which method will be most helpful to him?

   A) Construct an Euler circuit or minimal Eulerization.

   B) Construct a minimum-cost Hamiltonian circuit.

   C) Construct a minimum-cost spanning tree.

   D) Construct a vertex coloring in the fewest possible colors.

12. An engineer needs to connect computers on each floor of a building and then connect each floor to the server. Which method will be most helpful to her?

   A) Construct an Euler circuit or minimal Eulerization.

   B) Construct a minimum-cost Hamiltonian circuit.

   C) Construct a minimum-cost spanning tree.

   D) Construct a vertex coloring in the fewest possible colors.

# Scheduling and Bin Packing

## Worksheet 4

1. Consider the order-requirement digraph given below.

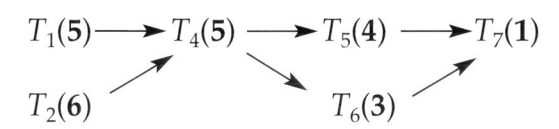

   $T_1(5) \longrightarrow T_4(5) \longrightarrow T_5(4) \longrightarrow T_7(1)$

   $T_2(6) \nearrow \qquad\qquad\searrow \qquad\qquad \nearrow$

   $\qquad\qquad\qquad\qquad T_6(3)$

   $T_3(4)$

   Use the list-processing algorithm with priority list $T_1, T_2, T_3, T_4, T_5, T_6, T_7$ and two processors to schedule the tasks.

2. For the order-requirement digraph given in Problem 1, use the list-processing algorithm with priority list $T_7, T_6, T_5, T_4, T_3, T_2, T_1$ and three processors to schedule the tasks.

3. For the order-requirement digraph given in Problem 1, use the critical-path scheduling algorithm and two processors to schedule the tasks.

4. Use the decreasing time-list algorithm to schedule the following independent tasks on two processors.

   $T_1(4), T_2(5), T_3(7), T_4(2), T_5(9), T_6(2), T_7(4), T_8(6)$

5. What is the best possible time in which nine independent tasks could be scheduled on three processors if the sum of all the times of the tasks is 75 minutes?

6. Find the critical path in the following order-requirement digraph.

$T_1(5) \longrightarrow T_2(4) \longrightarrow T_3(6)$

$T_4(5) \longrightarrow T_5(8)$

$T_6(7) \longrightarrow T_7(3) \longrightarrow T_8(3)$

7. Consider the following nine weights:

   5 lbs, 6 lbs, 3 lbs, 2 lbs, 6 lbs, 4 lbs, 7 lbs, 4 lbs, 3 lbs

   Use the first-fit bin-packing algorithm to pack the weights into bins which have a maximum capacity of 10 lbs.

8. Use the next-fit bin-packing algorithm to pack the weights given in Problem 6 into bins which have a maximum capacity of 10 lbs.

9. Use the worst-fit decreasing bin-packing algorithm to pack the weights given in Problem 6 into bins that have a maximum capacity of 10 lbs.

10. Two schedules for the same set of tasks on the same number of processors have the same completion time. Can one schedule have more idle time than the other?

61

# Sample Exam 4

1. For the order-requirement digraph given below and the priority list $T_1, T_2, T_3, T_4, T_5, T_6, T_7, T_8, T_9$, how much time would the schedule resulting from the list-processing algorithm on two processors take?

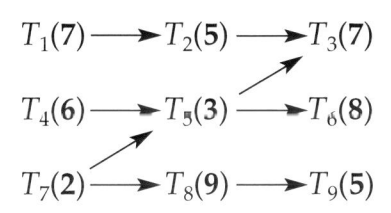

$$T_1(7) \longrightarrow T_2(5) \longrightarrow T_3(7)$$
$$T_4(6) \longrightarrow T_5(3) \longrightarrow T_6(8)$$
$$T_7(2) \longrightarrow T_8(9) \longrightarrow T_9(5)$$

A) 28

B) 33

C) 29

D) 34

2. For the order-requirement digraph given in Problem 1, how much time would the schedule resulting from the critical-path scheduling algorithm on two processors take?

A) 28

B) 33

C) 29

D) 34

3. How much time would the schedule resulting from the decreasing time-list algorithm applied to tasks of time

   3 min, 5 min, 6 min, 2 min, 8 min, 5 min, 4 min, and 3 min

   using two processors take?

   A) 16

   B) 17

   C) 18

   D) 19

4. A circus has 15 acts of various lengths which they want to efficiently fit into three segments separated by intermissions. The algorithm that would be useful here is

   A) the list-processing algorithm.

   B) the decreasing time-list algorithm.

   C) the first-fit algorithm.

   D) the worst-fit decreasing algorithm.

5. When using the list-processing algorithm, increasing the number of processors used will always reduce the completion time.

   A) True

   B) False

6. Consider the following weights:

   3 lbs, 5 lbs, 6 lbs, 2 lbs, 7 lbs, 2 lbs, 2 lbs, 5 lbs, 4 lbs.

   How many bins are used when using the next-fit algorithm on bins with a maximum capacity of 10 lbs to pack the given weights?

   A) 4

   B) 7

   C) 6

   D) 5

7. When using the first-fit algorithm on the weights and bins given in Problem 6, the third bin would contain

   A) two weights each being 5 lbs.

   B) one 7 lb weight.

   C) a 7 lb weight and a 3 lb weight.

   D) a 7 lb weight and a 2 lb weight.

8. When using the worst-fit algorithm on the weights and bins given in Problem 6, the third bin would contain

   A) two weights each being 5 lbs.

   B) one 7 lb weight.

   C) a 7 lb weight and a 3 lb weight.

   D) a 7 lb weight and a 2 lb weight.

9. When using the first-fit decreasing algorithm on the weights and bins given in Problem 6, the third bin would contain

   A) two weights each being 5 lbs.

   B) one 7 lb weight.

   C) a 7 lb weight and a 3 lb weight.

   D) a 7 lb weight and a 2 lb weight.

10. How long is the critical path given in the following order-requirement digraph?

    $T_1(8) \longrightarrow T_2(6) \longrightarrow T_3(5)$

    $T_4(7) \longrightarrow T_5(5) \longrightarrow T_6(8)$

    $T_7(6) \longrightarrow T_8(7) \longrightarrow T_9(4)$

    A) 20

    B) 21

    C) 22

    D) 23

11. Independent tasks of durations 12, 14, 8 and 6 minutes are scheduled on two machines using the decreasing-time-list algorithm. How long does the schedule require?

    A) 24 minutes
    B) 26 minutes
    C) 28 minutes
    D) 30 minutes

12. An order-requirement digraph requires 45 minutes when scheduled on 2 machines. When it is scheduled on three machines, it will require:

    A) shorter than 30 minutes
    C) at least 30 minutes
    B) exactly 30 minutes
    D) exactly 45 minutes

# DESCRIPTIVE STATISTICS

## WORKSHEET 5

1. In a survey, 900 student names were selected randomly from the 10,000 names obtained from the registrar. The 900 chosen students were then asked if they smoke. 200 responded yes, 700 responded no. What is the population in this study? What is the sample in this study? What is the sample proportion of those who responded yes?

2. To determine the proportion of people in a town who liked a particular cafe, a sample of people leaving the cafe were surveyed. It was found that 60 percent of them did like the cafe. Explain why there is bias in this study.

3. Define the terms placebo, experiment, and voluntary-response sample.

4. Below are the ages of 15 children at a playground.

    9 13 6 12 7 8 12 11 10 8 9 12 13 9 7

    Make a dotplot of this data.

5. Make a histogram of the data given in Problem 1 using widths of two years.

6. What is the median of the data given in Problem 1?

7. What are the first and third quartiles of the data given in Problem 1?

8. What is the five-number summary of the data given in Problem 1?

9. What is the mean of the data given in Problem 1?

10. What is the standard deviation of the data given in Problem 1?

11. A pizza shop collected data comparing the number of pizzas prepared (Y) and the number of people served (X). Below is the data.

| X | Y | X | Y |
|---|---|---|---|
| 10 | 8 | 22 | 13 |
| 50 | 22 | 44 | 25 |
| 30 | 19 | 60 | 32 |
| 40 | 23 | 28 | 18 |

Make a scatterplot of this data.

12. The pizza shop in Problem 8 has determined that the least-squares regression equation for the data is $Y = .5X + 3$. Predict how many pizzas will be needed for 110 people.

# Sample Exam 5

1. Consider the following data set:

   10, 30, 15, 20, 30, 20, 25, 20, 25, 20

   Which dotplot below represents this data?

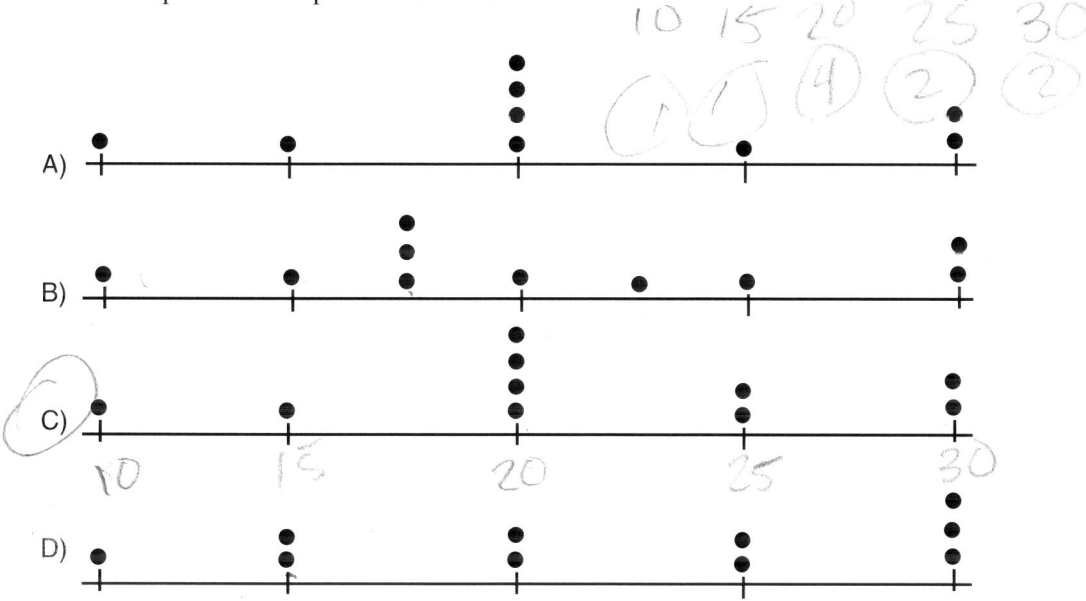

   A)

   B)

   C)

   D)

2. Find the median of the data given in Problem 1.

   A) 15

   B) 20

   C) 25

   D) 30

3. Find the third quartile of the data given in Problem 1.

   A) 25

   B) 27.5

   C) 22.5

   D) 20

4. Find the mean of the data given in Problem 1.

   A) 20

   B) 22.5

   C) 21.5

   D) 25

5. Find the standard deviation of the data given in Problem 1.

   A) 5.94

   B) 6.33

   C) 6.26

   D) 5.56

6. In a survey, 500 people are chosen at random from a group of 10,000 people. What is the set of 500 people called?

   A) The population.

   B) The sample.

   C) The control group.

   D) The experiment.

7. The histogram below gives the number of minutes that people spend in the shower each morning. How many people spend between four and six minutes in the shower each morning?

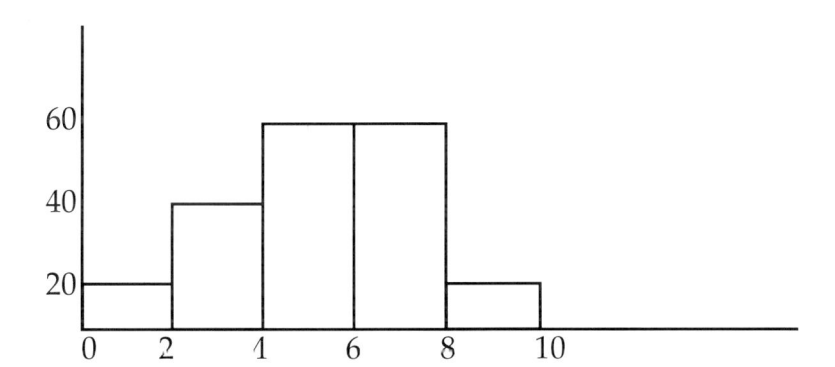

   A) 60

   B) 40

   C) 20

   D) 120

8. A fake medication will sometimes show signs of helping a patient. This is an example of

   A) bias.

   B) confounding variables.

   C) the placebo effect.

   D) variability.

9. In a survey of 2,000 people, it was found that those who smoke had more heart problems than those who don't smoke. This study cannot conclude that smoking causes heart problems because

   A) this study was not an experiment.

   B) this study had voluntary response.

   C) the study is not double-blind.

   D) of the placebo effect.

10. A company keeps data on the number of pounds of an order compared to the cost of the order. The least-squares regression line is given by $C = 20 + 30p$. Use this to estimate the cost of a 50 pound order.

    A) 1500

    B) 1530

    C) 1520

    D) 1550

11. The last of the five numbers of the five-number summary of the data given in Problem 1 is

   A) 30

   B) 10

   C) 15

   D) 25

12. Scores on an examination have a mean of sixty and a standard deviation of three. What is the variance of scores on this exam?

   A) 3

   B) 7

   C) 5

   D) 9

# PROBABILITY, NORMAL CURVES

## WORKSHEET 6

1. If two fair dice are rolled, what is the probability of getting a sum less than six?

2. Two coins are tossed and a die is rolled. What is the sample space of this experiment?

3. What is the probability of getting at least one head on a coin and a number less than three on the die in the experiment given in Problem 2?

4. A license-plate code consists of three letters followed by three digits. What is the probability that a randomly chosen license plate has no letters the same and no digits the same?

5. Below is a probability model of TVs owned by a household in the US. What is the mean number of TVs owned?

| No. of TVs | 0 | 1 | 2 | 3 | 4 | 5 |
|---|---|---|---|---|---|---|
| Probability | .10 | .35 | .25 | .15 | .10 | .05 |

6. The distribution of the heights of adult women is normally distributed. It has a mean of 63 inches and a standard deviation of 2.2 inches. What is the upper quartile of this distribution?

7. Referring to Problem 6, between what two values does the middle 68% of the heights of adult women lie?

8. A jar contains 30% black beans and 70% red beans. A sample of 500 beans is selected and it is found that 28% of the beans in the sample are black. Which of the numbers 30% or 28% is the parameter and which is the statistic?

9. In Problem 7, between which values does the middle 50% of heights lie?

10. Stipends of research students are normally distributed with a mean of $16,000 and a standard deviation of $1,800. Within what range do the top 2.5% of stipends lie?

95

# Sample Exam 6

1. There are 12 red balls and 24 blue balls in a basket. If a ball is pulled at random from the basket, what is the probability that it is red?

    A) 1/2

    B) 1/36

    C) 12/100

    D) 1/3

2. A game has four outcomes W, X, Y and Z with probabilities P(W) = 0.2, P(X) = 0.3, P(Y) = 0.1, and P(Z) = 0.4. A player pays $2 if W occurs, pays $2 if X occurs, receives $1 if Y occurs, and receives $3 if Z occurs. What is the mean value of this model?

    A) 0.3

    B) 0.4

    C) 0.2

    D) 0.1

97

3. Two dice are rolled and two coins are tossed. How many outcomes are in the sample space of this experiment?

   A) 20

   B) 40

   C) 124

   D) 144

4. A password consists of three digits. What is the probability that a randomly selected password has three different odd digits?

   A) 1/8

   B) 8/39

   C) 3/50

   D) 3/1000

5. Five coins are flipped and the number of coins that landed on tails is recorded. Are the outcomes in this sample space all equally likely?

   A) Yes

   B) No

6. A game has three outcomes A, B and C. The probability of A is 0.25 and the probability of B is 0.6. What is the probability of C?

    A) 0.05

    B) 0.85

    C) 0.25

    D) 0.15

7. 30% of Republican voters are in favor of stronger gun control laws. 1,000 Republican voters will be surveyed with the percent in favor of stronger gun control laws being the statistic of interest. What is the standard deviation of the sampling distribution of this statistic?

    A) 1.45%

    B) 2.20%

    C) 1.98%

    D) 2.53%

8. The mean length of time that children spend watching TV is 20 hours per week with a standard deviation of 1.5 hours. Assuming that this distribution is normal, what percent of children watch TV between 15.5 and 24.5 hours per week?

    A) 60%

    B) 95%

    C) 68%

    D) 99.7%

9. What is the third quartile of the distribution give in Problem 8?

   A) 20.92

   B) 21.01

   C) 22.67

   D) 23.44

10. A random sample of 1,200 teenagers were asked if they smoke cigarettes and 300 of them answered yes. What is the population in this study?

    A) The 300 teenagers who said they smoke.

    B) The 1,200 teenagers interviewed.

    C) All teenagers who smoke.

    D) All teenagers.

11. A poll of 2,000 people in a town finds that 22% are in favor of building a new golf course. In this study 22% is a

    A) standard deviation.

    B) parameter.

    C) statistic.

    D) mean.

12. In a certain town, family incomes are distributed normally with mean $34,000 and standard deviation $6,000. What percentage of incomes do you expect to lie between $30,000 and $40,000?

    A) 50%  B) 59%  C) 65%  D) 95%

# Sampling Distributions, Confidence Intervals, Code Numbers

## Worksheet 7

1. Suppose you need to halve the size of your confidence interval. What change should you make to your sample size?

2. A poll of 2,000 residents finds that 1,140 favor a recycling program. What is the sample proportion?

3. In a certain town, family incomes are distributed normally with mean $38,000 and standard deviation $8,000. A sample of 16 families is surveyed. What is the 95% confidence interval for the mean income of this sample?

4. In a survey of 900 people, it is found that 665 favor a candidate's platform. Find a 95% confidence interval for the real proportion of people who favor the candidate's platform.

5. What check digit should be appended to the Codabar number 172837452637123?

6. What check digit should be appended to the ISBN number 0-2645-4376?

7. What check digit should be appended to the Zip + 4 code 47634-0632?

8. A motorist's Illinois driver's license number is P110-3456-1095. Is the motorist male or female? When is the motorist's birthday?

9. Suppose a U.S. Postal money order is received as 364839#8425 where the seventh digit # cannot be read. What is the seventh digit?

10. Suppose a Codabar number is received as 283746#098572311 where the seventh digit # cannot be read. What is the seventh digit?

# Sample Exam 7

1. A poll of 60 students found that 20% were in favor of raising parking fees to pave two new parking lots. The standard deviation of this poll is about 5.20%. What would be the standard deviation if the same result were obtained from a sample of 240 students?

   A) 0.65%  B) 1.30%  C) 2.60%  D) 5.20%

2. A random sample of 1,200 teenagers were asked if they smoke cigarettes and 25% of them answered yes. Find a 95% confidence interval for the percent of all teenagers who smoke cigarettes.

   A) 23.75% – 26.25%  C) 22.5% - 27.5%

   B) 24.15% – 25.85%  D) 21% - 29%

113

3. A poll of 1,600 Marion residents found that 64% were in favor of higher taxes to support a baseball stadium. Find a 95% confidence interval for the proportion of the city's residents who are in favor.

    A) 60.4% to 67.6%

    B) 61.6% to 66.4%

    C) 62.8% to 65.2%

    D) 63.4% to 64.6%

4. A person has Illinois driver's license number H123-3217-5322. This person's birth year is in the range:

    A) 1965 – 1971

    B) 1972 – 1978

    C) 1979 – 1985

    D) 1986 – 1992

5. The person with Illinois driver's license number given in Problem 6 is a male.

    A) True

    B) False

6. The person with Illinois driver's license number given in Problem 6 was born in what month?

   A) July

   B) June

   C) October

   D) November

7. The error-detecting mechanism used for U.S. Postal money order cannot detect the mistake of replacing a 9 with a 0.

   A) True

   B) False

8. What check digit should be appended to the Codabar number 152737485924364?

   A) 5

   B) 6

   C) 7

   D) 8

9. What check digit should be appended to the ISBN number 0-1435-7243?

   A) 1				C) 5

   B) 3				D) 7

10. What check digit should be appended to the U.S. Postal money order number 2516242513?

    A) 1			C) 3

    B) 2			D) 4

11. Suppose a Zip + 4 code number is received as 4251#-2263-2 where the fifth digit # cannot be read. What is the fifth digit?

    A) 1			C) 3

    B) 2			D) 4

12. The error-detecting mechanism used for ISBN numbers can detect all single position errors.

    A) True

    B) False

# Error-Correcting Codes

## Worksheet 8

1. Add the binary sequences 1001010 and 1010010.

2. Determine the code words of the messages 1110 and 0101 using the Venn-diagram method.

3. What is the distance between the words 10111001 and 11100101?

4. What is the binary sum of the words 10001110 and 01011101?

5. What are the weights of the two words given in Problem 4?

6. Decode the received words 1001011 and 1110011 using the nearest-neighbor Venn-diagram method of decoding.

7. Consider the code $C = \{1111000, 1100110, 1010101, 0011110, 1010101, 0101101, 0110010, 0000000\}$. Is C a linear code?

8. What is the weight and minimum distance of the code C given in Problem 7.

9. What is the error-correcting and error-detecting capability of the code C given in Problem 7?

10. What is the error-correcting and error-detecting capability of the binary code obtained by appending to each message word $a_1a_2a_3a_4$ the parity check digits $a_5 = a_1 + a_2 + a_3$, $a_6 = a_1 + a_2 + a_4$, $a_7 = a_1 + a_3 + a_4$, $a_8 = a_2 + a_3 + a_4$

11. For the code given in Problem 10, decode the received words 11111110 and 00011100 using the nearest-neighbor method.

# SAMPLE EXAM 8

1. Add the binary sequences 1001010 and 1010010. How many 1s digits are in the sum?

   A) 3      B) 1      C) 4      D) 2

2. Given the message 0011, use the Venn-diagram method to determine the associated code word.

   A) 0011100      C) 1000100

   B) 0011101      D) 0011011

3. Find the distance between the words 010110 and 011100.

   A) 1      C) 3

   B) 2      D) 4

4. What is the weight of the binary sum of the words 1110010 and 0011010?

    A) 1
    
    B) 2
    
    C) 3
    
    D) 4

5. Decode the received word 0011010 using the nearest-neighbor Venn-diagram method.

    A) 0001
    
    B) 0011
    
    C) 0111
    
    D) 1011

6. The binary sum of two words of the same length and of even weight will always have even weight.

    A) True
    
    B) False

131

7. Consider the code C = {0000, 1100, 1011, 1110}. What is the weight of this code?

   A) 0
   B) 1
   C) 2
   D) 3

8. What is the minimum distance of the code C given in Problem 7?

   A) 0
   B) 1
   C) 2
   D) 3

9. If a code has minimum distance six, how many errors can it detect?

   A) 2
   B) 3
   C) 4
   D) 5

10. If a code has minimum distance six, how many errors can it correct?

    A) 2
    
    B) 3
    
    C) 4
    
    D) 5

11. Consider the binary code obtained by appending to each message $a_1a_2a_3$ the parity check digits $a_4 = a_1 + a_3$ and $a_5 = a_2 + a_3$. Which of the following are code words?

    1) 11100
    
    2) 00110
    
    A) (1) only
    
    B) (2) only
    
    C) both (1) and (2)
    
    D) neither (1) nor (2)

12. How many errors can the code given in Problem 11 correct?

    A) 0
    
    B) 1
    
    C) 2
    
    D) 3

# VOTING AND GEOMETRIC GROWTH

## WORKSHEET 9

1. Why is majority rule not a good way to choose between three or more alternatives?

2. How many ways is it possible to rank ten candidates?

3. Consider the preference schedule given below. Using the majority rule method of voting, which candidate wins the election?

|            | 8 | 6 | 5 | 2 |
|------------|---|---|---|---|
| 1st choice | A | C | B | D |
| 2nd choice | D | A | C | A |
| 3rd choice | B | B | D | B |
| 4th choice | C | D | A | C |

4. Using the preference schedule given in Problem 3 and the plurality method of voting, which candidate wins the election?

5. Using the preference schedule given in Problem 3 and the Hare method of voting, which candidate wins the election?

6. Using the preference schedule given in Problem 3 and the Condorcet method of voting, which candidate wins the election?

7. Using the preference schedule given in Problem 3 and the Borda count (7, 5, 3, 1), which candidate wins the election?

8. Seven voters vote by approval voting on five candidates, A, B, C, D, and E. The results are listed below with * indicating approval. Which candidate wins?

|   | 1 | 2 | 3 | 4 | 5 | 6 | 7 |
|---|---|---|---|---|---|---|---|
| A | * |   | * | * | * |   |   |
| B |   | * | * |   | * |   |   |
| C | * |   | * |   | * | * | * |
| D | * |   | * |   |   | * |   |
| E | * |   |   | * | * |   |   |

9. If $15,000 is invested at 9% compounded annually, what is the balance after five years?

10. If $15,000 is invested at 9% compounded daily, what is the balance after five years?

11. A colony of birds had 2,000 members on January 1, 2006. The colony expands at 5% per year. How many members do you expect there to be on January 1, 2008?

141

# Sample Exam 9

1. In a majority rule election with thirty voters and three candidates, how many votes does a winner need?

   A) 11

   B) 10

   C) 15

   D) 16

2. Consider the following preference schedule. Which candidate wins using a Borda count (4, 3, 2, 1)?

   |  | 4 | 3 | 3 |
   |---|---|---|---|
   | 1st choice | B | C | D |
   | 2nd choice | C | A | A |
   | 3rd choice | D | D | B |
   | 4th choice | A | B | C |

   A) A

   B) B

   C) C

   D) D

3. Can the three voters in the last column of the preference schedule given in Problem 2 vote strategically for a result that is more favorable for them in that election?

   A) Yes, switch A and B.

   B) Yes, switch C and D.

   C) Yes, switch A and D.

   D) No.

4. Which candidate wins the election of Problem 2 if the plurality method of voting is used.

   A) A

   B) B

   C) C

   D) D

5. Consider the following preference schedule. Which candidate wins if the Hare voting method is used?

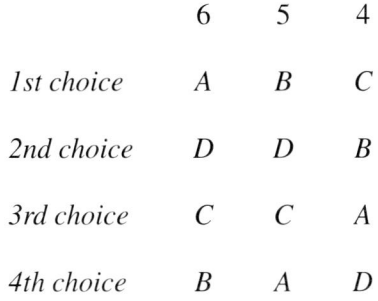

|  | 6 | 5 | 4 |
|---|---|---|---|
| 1st choice | A | B | C |
| 2nd choice | D | D | B |
| 3rd choice | C | C | A |
| 4th choice | B | A | D |

A) A
B) B
C) C
D) D

6. Which candidate wins the election of Problem 5 if the sequential-pairwise method of voting with agenda A, B, C, D is used?

A) A
B) B
C) C
D) D

7. For the preference schedule given in Problem 5, is there a Condorcet winner?

   A) Yes, A.
   
   C) Yes, C.
   
   B) Yes, B.
   
   D) No.

8. Nine voters vote by approval voting on five candidates, A, B, C, D, and E. The results are listed below with * indicating approval. Which candidate wins?

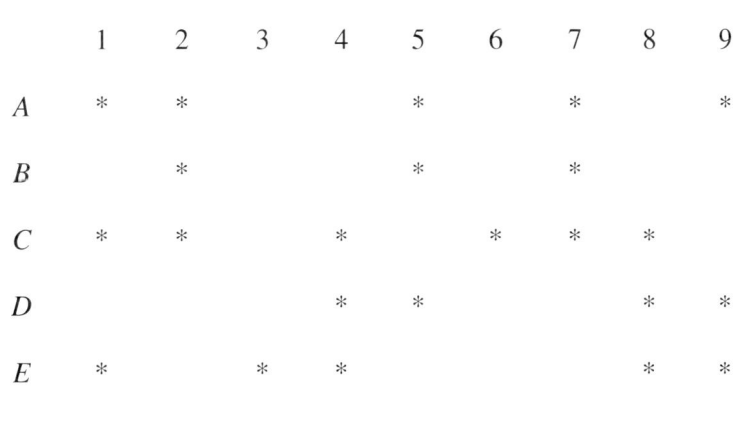

   |   | 1 | 2 | 3 | 4 | 5 | 6 | 7 | 8 | 9 |
   |---|---|---|---|---|---|---|---|---|---|
   | A | * | * |   | * |   | * |   |   | * |
   | B |   | * |   |   | * |   | * |   |   |
   | C | * | * |   | * |   | * | * | * |   |
   | D |   |   |   | * | * |   |   | * | * |
   | E | * |   | * | * |   |   |   | * | * |

   A) A
   
   C) C
   
   B) B
   
   D) E

9. A majority-rule winner will always win an election using the plurality voting method.

   A) True
   
   B) False

10. $2,000 is invested at 6% interest compounded quarterly. What is the balance after two years?

   A) $2,240

   B) $2,247

   C) $2,253

   D) $2,262

11. At noon Monday there are 10,000,000 bacteria present in a colony. At noon on Wednesday there are 12,100,000 present. How many do you expect at noon on Thursday?

   A) 11,000,000

   B) 14,641,000

   C) 24,200,000

   D) 13,310,000

12. An artificial element has a half-life of three hours. If 640gm is present at noon, how much is present at 9PM?

   A) 40gm

   B) 80gm

   C) 160gm

   D) 480gm

# REVIEW

## FINAL WORKSHEET

1. Show that the graph given below has no Euler circuit.

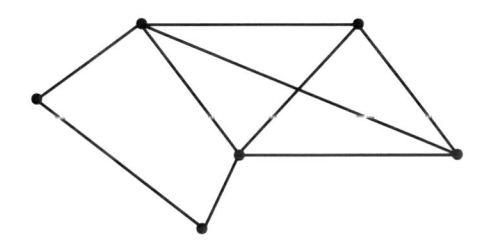

2. Find an Euler circuit on the graph given below.

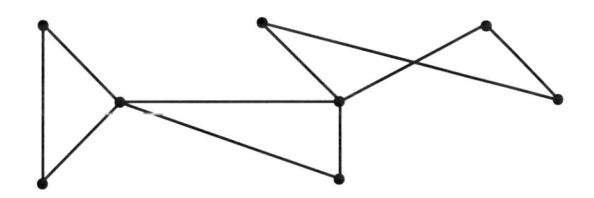

3. Eulerize the graph given in Problem 1.

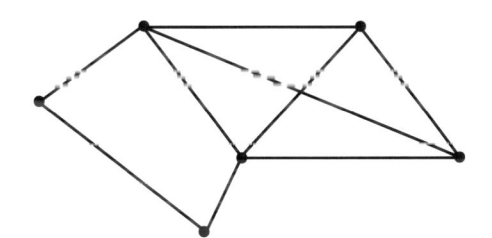

4. For the graph given below, use the sorted-edges algorithm to find a route for a traveling salesman.

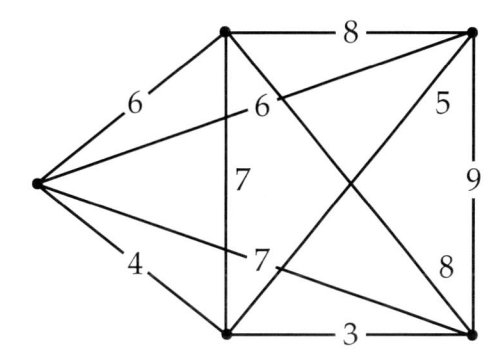

5. For the graph given in Problem 4, use the nearest-neighbor algorithm starting at the far left vertex to find a route for a traveling salesman.

6. For the graph given in Problem 4, use Kruskal's algorithm to find a minimal-cost spanning tree.

155

7. For the order-requirement digraph given below and the priority list $T_1, T_2, T_3, T_4, T_5, T_6, T_7, T_8$, use the list-processing algorithm to construct a schedule for two processors.

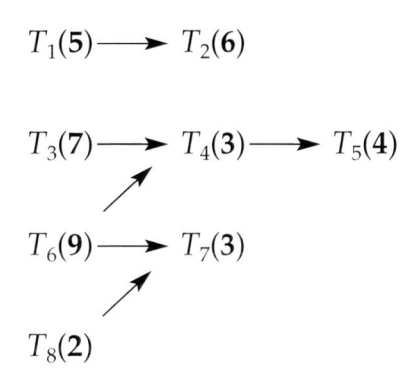

$T_1(5) \longrightarrow T_2(6)$

$T_3(7) \longrightarrow T_4(3) \longrightarrow T_5(4)$

$T_6(9) \longrightarrow T_7(3)$

$T_8(2)$

8. Find the chromatic number of the following graph:

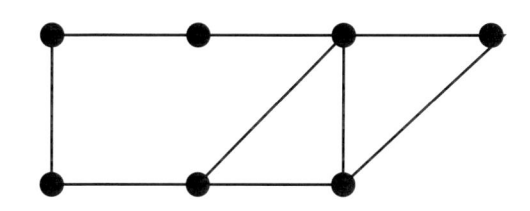

9. Use the first-fit decreasing bin-packing algorithm to pack the following weights into bins of capacity 9 lbs.

   5 lbs, 7 lbs, 3 lbs, 2 lbs, 6 lbs, 4 lbs, 1 lb, 7 lbs, 5 lbs

10. Find the first quartile and standard deviation for the following set of data.

    3, 12, 4, 5, 10, 10, 5, 5, 5, 2, 6, 7, 1

11. A license code consists of four digits. What is the probability that a randomly chosen license has a code with all four digits the same?

12. Two dice are rolled. If the sum is 5, 6, 7, or 8, you receive $1.00. Otherwise, you pay $1.00. What is the mean value of one trial of this game?

13. Convert the decimal number 107 to binary notation.

14. A survey finds that 600 of 2,000 students are in favor of a restricted smoking area in the student center. Find a 95% confidence interval for the true proportion of students who favor the restricted smoking area.

15. If a man has Illinois driver's license number T 123 6547 1100, when was he born?

16. The ISBN number 0-1536-362#-6 has been received where the 9th digit # cannot be read. What is this 9th digit?

17. The shelf life of an aspirin is normally distributed with a mean of 1.25 years and a standard deviation of 0.5 years. Between what two values does 99.7% of all aspirin shelf lives lie?

18. Using the nearest-neighbor Venn-diagram method, decode the received word 0001011.

19. Consider the binary code obtained by appending to each word $b_1b_2b_3b_4$ the parity-check digits $b_5 = b_1 + b_2 + b_3$, $b_6 = b_1 + b_3 + b_4$, and $b_7 = b_2 + b_3 + b_4$. How many errors can this code detect and how many can it correct?

20. Suppose $4,000 is deposited at 8% compounded daily. What is the balance after three years?

21. Consider the following preference schedule. Which candidate wins the election using the Hare system of voting?

|            | 10 | 9 | 8 |
|------------|----|---|---|
| 1st choice | A  | D | E |
| 2nd choice | C  | E | C |
| 3rd choice | D  | C | B |
| 4th choice | E  | A | A |
| 5th choice | B  | B | D |

22. If a Borda count (5, 4, 3, 2, 1) is used for the election in Problem 21, which candidate wins?

23. Can the eight voters in the last column of the election in Problem 22 vote strategically to obtain a more favorable outcome?

24. Who wins the election of Problem 22 if sequential-pairwise voting is used with the agenda A, B, C, D, E?

165

# SAMPLE FINAL EXAM

1. For the following graph, which of the following statements is true?

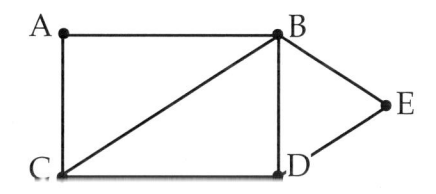

   1) ABEDCA is a Hamiltonian circuit.

   2) CABDEBC is an Euler circuit.

   A) (1) only

   B) (2) only

   C) both (1) and (2)

   D) neither (1) nor (2)

2. Every graph with an Euler circuit is connected.

   A) True

   B) False

167

3. Which of the following graphs are trees?

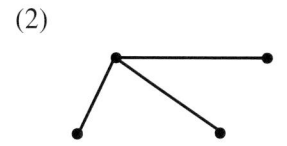

(1)

(2)

A) (1) only

B) (2) only

C) both (1) and (2)

D) neither (1) nor (2)

4. If all of the vertices of a connected graph have even valence, then the graph

A) has an Euler circuit and a Hamiltonian circuit.

B) has an Euler circuit but may not have a Hamiltonian circuit.

C) has a Hamiltonian circuit but may not have an Euler circuit.

D) has neither an Euler circuit nor a Hamiltonian circuit.

5. A minimal-cost spanning tree in the following graph has what cost?

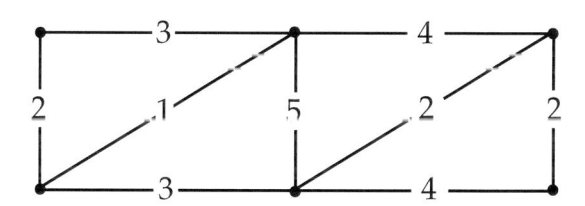

A) 10

B) 11

C) 12

D) 13

169

`6. Consider the following graph. What is the cost of the Hamiltonian circuit obtained by using the nearest-neighbor algorithm starting at vertex A?

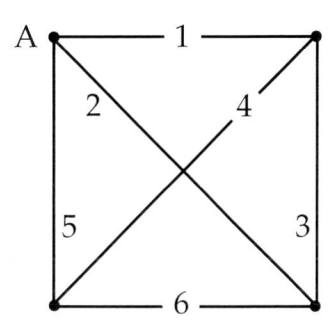

A) 12

B) 13

C) 14

D) 15

7. On the graph given in Problem 6, what is the cost of the Hamiltonian circuit obtained by applying the sorted-edges algorithm?

A) 12

B) 13

C) 14

D) 15

8. What is the chromatic number of this graph?

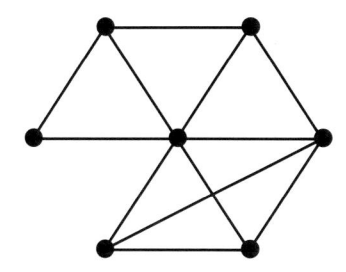

A) 7   B) 6   C) 4   D) 2

9. Using the decreasing-time-list algorithm on tasks of length 6, 3, 3, 4, 5, and 8 hours, how long does it take for two processors to finish the tasks?

   A) 13 hours
   B) 14 hours
   C) 15 hours
   D) 16 hours

10. What is the length of the critical path in the following order-requirement digraph?

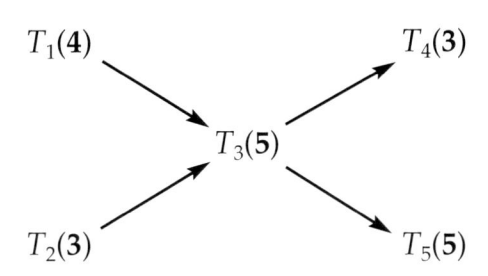

$T_1(4)$         $T_4(3)$

$T_3(5)$

$T_2(3)$         $T_5(5)$

A) 12

B) 13

C) 14

D) 15

11. How long does it take to finish the tasks when applying the list-processing algorithm to the order-requirement digraph of Problem 9 using two processors and the priority list $T_1, T_2, T_3, T_4, T_5$?

A) 13

B) 14

C) 15

D) 16

12. When using the first-fit bin-packing algorithm on bins of capacity 10 lbs and weights 3 lbs, 5 lbs, 7 lbs, 4 lbs, 2 lbs, 5 lbs, and 1 lb, what weights are contained in the second bin?

    A) a 7 lb weight and a 1 lb weight

    B) a 7 lb weight and a 2 lb weight

    C) a 7 lb weight and a 3 lb weight

    D) a 7 lb weight only

13. Confounding occurs in an experiment when

    A) the sample is not random.

    B) the effects of one variable cannot be distinguished from the effects of another variable.

    C) the placebo effect is present.

    D) the experiment is not double-blind.

14. The daily high temperatures recorded during a week in Canada are: 6, 6, 5, 9, 11, 8, and 4. What is the mean of these temperatures?

    A) 9                    C) 8

    B) 7                    D) 6

15. The median of the temperatures given in Problem 14 is

    A) 9
    B) 7
    C) 8
    D) 6

16. The 3rd quartile of the temperatures given in Problem 14 is

    A) 9
    B) 7
    C) 8
    D) 6

17. If a number from 1 to 30 is chosen at random, what is the probability that the number is larger than 25?

    A) 1/3
    B) 1/8
    C) 1/5
    D) 1/6

18. Standard deviation is a measure of

    A) center.

    B) skewness.

    C) symmetry.

    D) spread.

19. An exam is taken by 500 students. The scores on the exam are normally distributed with a mean of $\mu = 64$ and standard deviation $\sigma = 10$. What percentage of students scored between 44 and 84 on the exam?

    A) 67%

    B) 95%

    C) 68%

    D) 99.7%

20. In a standardized test, the mean score is 500 points. Scores approximate a normal distribution with standard deviation 100. It is decided that 25% will fail. What is the smallest (whole number) score that will receive a passing grade?

    A) 375

    B) 477

    C) 400

    D) 434

21. 400 adults were asked if they jog and 20% answered yes. Find a 95% confidence interval for the percent of all adults who jog.

    A) 18–22

    B) 19–21

    C) 16–24

    D) 14–26

22. How many numbers between 2 and 1,000 start with a 1 and end with a 7?

    A) 8

    B) 9

    C) 10

    D) 11

23. A person has Illinois driver's license number P-110-3455-6366. This person's age is in what range?

    A) 20–30

    B) 30–40

    C) 40–50

    D) 50–60

24. The person in Problem 23 is a male.

    A) True

    B) False

25. What check-digit should be appended to the ISBN number 0-121-56213?

    A) 1          C) 4

    B) 9          D) 3

26. Use the Venn-diagram method to encode the message 0001.

    A) 0001011          C) 0001001

    B) 0001000          D) 0001110

27. What is the minimum distance of the code {1100, 1010, 1001, 1101}?

    A) 1

    B) 2

    C) 3

    D) 4

28. The population of a country doubles every 15 years. If the population is currently 3 million people, in how many years will it reach 12 million people?

    A) 40 years

    B) 45 years

    C) 30 years

    D) 60 years

29. Add the binary sequences 1001010 and 1011010. How many 1s digits are in the sum?

    A) 1
    B) 2
    C) 3
    D) 4

30. The preference schedule for 20 voters is shown below. Which candidate wins if the Hare system of voting is used?

|  | 5 | 5 | 4 | 6 |
|---|---|---|---|---|
| 1st choice | A | C | A | B |
| 2nd choice | B | B | C | A |
| 3rd choice | C | A | B | C |

A) A

B) B

C) C

D) there is no winner

31. Which candidate wins the election of Problem 30 if majority rule voting is used?

A) A

B) B

C) C

D) there is no winner

32. Which candidate wins the election of Problem 30 if a Borda count (3, 2, 1) is used?

A) A

B) B

C) C

D) there is no winner

33. Is there a Condorcet winner for the preference schedule given in Problem 30?

   A) Yes

   B) No

34. Suppose $500 is invested at 8% compounded quarterly. What is the balance after 2 years?

   A) $541.21

   B) $585.82

   C) $583.20

   D) $580

35. You borrow $4,000 at 6% annual interest, compounded every 3 months, and pay it back 6 months later. How much must you pay (rounded to the nearest dollar)?

   A) $4,000

   B) $4,240

   C) $4,121

   D) $4,244

36. On January 1 2004 there were 10,000 white rats in a colony being bred for laboratory use. On January 1 2007 there are 13,310 present. How many do you expect on January 1 2008?

    A) 16,050

    B) 14,641

    C) 26,620

    D) 13,310

# ANSWERS

## SAMPLE EXAMS

| Exam 1 | Exam 2 | Exam 3 | Exam 4 |
|---|---|---|---|
| 1-*B* | 1-*C* | 1-*D* | 1-*A* |
| 2-*A* | 2-*A* | 2-*A* | 2-*A* |
| 3-*B* | 3-*B* | 3-*B* | 3-*C* |
| 4-*A* | 4-*B* | 4-*A* | 4-*B* |
| 5-*B* | 5-*A* | 5-*B* | 5-*B* |
| 6-*C* | 6-*C* | 6-*A* | 6-*D* |
| 7-*B* | 7-*C* | 7-*C* | 7-*B* |
| 8-*A* | 8-*D* | 8-*B* | 8-*D* |
| 9-*C* | 9-*B* | 9-*A* | 9-*A* |
| 10-*B* | 10-*A* | 10-*A* | 10-*A* |
| 11-*A* | 11-*B* | 11-*D* | 11-*B* |
| 12-*C* | 12-*D* | 12-*C* | 12-*C* |

| Exam 5 | Exam 6 | Exam 7 | Exam 8 |
|---|---|---|---|
| 1-*C* | 1-*D* | 1-*C* | 1-*D* |
| 2-*B* | 2-*A* | 2-*C* | 2-*A* |
| 3-*A* | 3-*D* | 3-*B* | 3-*B* |
| 4-*C* | 4-*C* | 4-*B* | 4-*C* |
| 5-*A* | 5-*B* | 5-*A* | 5-*D* |
| 6 *B* | 6 *D* | 6-*D* | 6-*A* |
| 7-*A* | 7-*A* | 7-*A* | 7-*C* |
| 8-*C* | 8-*D* | 8-*A* | 8-*B* |
| 9-*A* | 9-*B* | 9-*A* | 9-*D* |
| 10-*C* | 10-*C* | 10-*D* | 10-*A* |
| 11-*A* | 11-*C* | 11-*C* | 11-*A* |
| 12-*D* | 12-*B* | 12-*A* | 12-*A* |

**Exam 9**

1-*D*
2-*C*
3-*A*
4-*B*
5-*B*
6-*D*
7-*D*
8-*C*
9-*A*
10-*C*
11-*D*
12-*B*

**Final Exam**

1-*A*
2-*A*
3-*C*
4-*B*
5-*A*
6-*D*
7-*B*
8-*C*
9-*C*
10-*C*
11-*B*
12-*A*
13-*B*
14-*B*
15-*D*
16-*A*
17-*D*
18-*D*
19-*B*
20-*D*
21-*C*
22-*D*
23-*D*
24-*A*
25-*A*
26-*A*
27-*A*
28-*C*
29-*C*
30-*B*
31-*D*
32-*A*
33-*A*
34-*B*
35-*C*
36-*B*

# Some Lecture Notes

1. Handshakes at a Party   201

2. Finding All Hamilton Circuits   203

3. Designing Experiments Using Latin Squares   205

4. About Standard Deviations   207

5. Confidence Intervals of Polls   209

6. The Illinois Driver's License   211

7. Everyone Wins   215

8. The Generalized Hare Method   217

9. Radioactive Decay   219

# 1 HANDSHAKES AT A PARTY

*A puzzle* Eight people — four married couples, husband and wife — attend a dinner party. Various introductions are made; everybody shakes hands (once!) with some of the others. Of course, no one shakes hands with his/her spouse.

After a while, the hostess asks every other person in the room, "how many times did you shake hands tonight?" It turns out that no two people shook hands the same number of times. (The hostess isn't included of course.)

How many times had the host (the hostess' spouse) shaken hands?

**Remark** We shall use graph theory to represent the elements of the problem. The question then becomes one of finding the valences of certain vertices. At first it seems that there is not enough information to solve the puzzle.

***The solution*** For convenience, let us write W, X, Y, Z, w, x, y, z for the eight people; W and w are one couple, X and x another, and so on. We represent the party by a graph with people as vertices and handshakes as edges. The number of handshakes is the valence of the vertex.

No one shook hands with more than six others (there are only six other people other than the two members of any couple). So the numbers of times people (other than the hostess) shook hands are seven different whole numbers, ranging from 0 to 6. It follows that each of 0, 1, 2, 3, 4, 5, 6 occurs exactly once among those other than the hostess.

So somebody shook hands 6 times. Let's call this person X. He or she did not shake hands with his/her spouse x, so he/she shook with everybody else. These handshakes contribute the edges shown in the following figure:

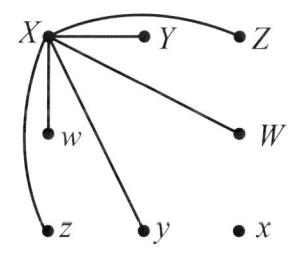

Somebody shook hands 0 times. Looking at the figure, every vertex received at least one edge, except for x. (There might be further edges, not yet shown.) So x is the person with no edges (no handshakes). No matter what happens, you will never add an edge touching x.

Somebody shook hands exactly five times. It was neither X nor x. Suppose it was Y. There is already one edge touching Y, the edge XY, so there are exactly four more edges touching Y. None of the new edges touch x or y (remember, people don't shake with spouses). This implies that y is the only person to receive exactly only one: x has 0, and every other vertex has at least 2. The figure looks like the following (possibly with more edges):

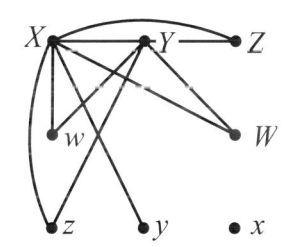

201

Next, someone shook hands exactly four times. It cannot be *X, x, Y* or *y*; say it is *Z*. There are already two edges touching *Z*, and *Z* is not joined to *z*. So *Z* must be joined to *W* and *w*. We have:

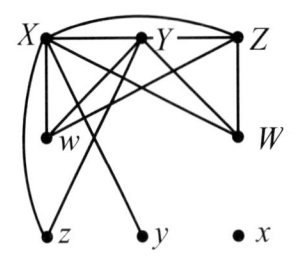

By now we have constructed a graph in which there are two vertices of valence 3 (*W* and *w*), and each of 0, 1, 2, 3, 4, 5 occurs as a valence once. One of the two vertices of degree 3 must be the hostess (this is the only possible way to get a repeat), and the other is her spouse. So the host shook hands three times.

# 2 FINDING ALL HAMILTON CIRCUITS

Suppose we want to find all Hamilton circuits in a graph $G$. We construct a new graph, a tree, which we shall call $T$.

Select any vertex of $G$ as a starting point; call it $V$ say. Construct the first vertex (the *root*) of the tree; label it with $V$. For every edge of $G$ that contains $V$, draw an edge of the tree and label it with the corresponding vertex of $G$.

We illustrate the method with the graph

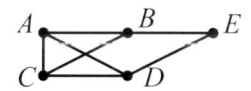

We can start anywhere: suppose we begin with vertex $A$. In the graph there are three vertices joined to $A$, namely $B$, $C$ and $D$. So we start three branches, and label them as shown.

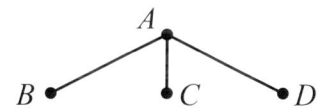

From each new vertex, draw an edge of the tree corresponding to each edge of $G$ that could be an edge of the Hamilton circuit (that is, it would not duplicate an edge of $G$ or complete a small circuit in $G$).

In the example, a branch will go from $B$ to vertices labeled $C$ and $E$, but not to $A$ (duplication) or $D$ (no edge in $G$). Similarly there are two branches from each of $C$ and $D$.

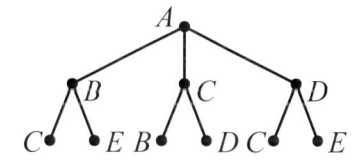

We continue in this way from each new vertex of the tree. For example, the path $A$, $B$, $E$ can extend only to $D$, because $B$ would be a repeated edge and $E$ is not connected to $A$ or $C$. The path $A$, $C$, $D$ can extend only to $E$, because $C$ would be a repeated edge and $A$ would form a small circuit.

As we continue, some of the branches cannot be continued. After $A$, $B$, $C$, $D$, $E$ there is no possible continuation, because $EA$ is not an edge of $G$. Others, like $A$, $C$, $B$, $E$, $D$, $A$ can be completed: the branch contains every vertex of $G$ and finishes back at $A$ — not a *small* circuit, because all the vertices are there. The final tree is

203

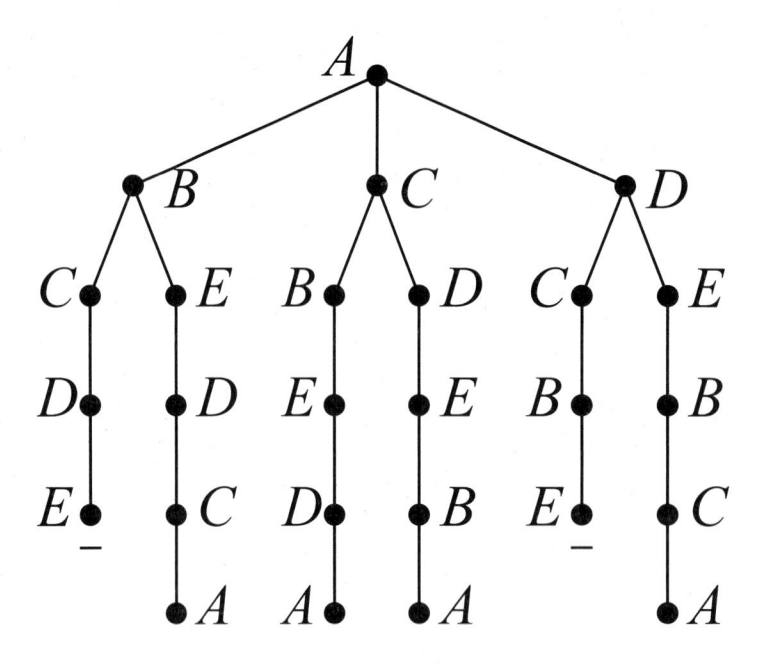

When the process is finished, the completed branches are the Hamilton cycles. Each cycle will occur twice, once in each direction.

So the example has two Hamilton cycles: *ABEDCA* and *ACBEDA*.

# 3 DESIGNING EXPERIMENTS USING LATIN SQUARES

*Experimental Designs* The combination of several experimental variables can make experiments prohibitively large and expensive. However, combinatorics, the mathematical study of arrangements, can sometimes provide clever schemes of treatments that help hold down the size and cost of experiments. These arrangements are called *experimental designs*. We will illustrate the idea by a comparison of motor oils.

*The Problem* Some makers of motor oil claim that using their product improves gasoline mileage in cars. Suppose you wish to test this claim by comparing the effect of four different oils on gas mileage. However, because the car model and the driver's habits greatly influence the mileage obtained, the effect of an oil may vary from car to car and from driver to driver. To obtain results of general interest, you must compare the oils in several different cars and with several different drivers. If you choose 4 car models and 4 drivers, there are 16 car-driver combinations. The type of car and driving habits are so influential that you must consider each of these 16 combinations as a separate experimental unit. If each of the 4 oils is used in each unit, 4 × 16, or 64, test drives are needed.

*The Design* Combinatorics provides a way to test the effects of cars, drivers, and oils that requires only 16 test drives. Auppose the cars are denoted $X_1$, $X_2$, $X_3$, $X_4$, and the drivers are $Y_1$, $Y_2$, $Y_3$, $Y_4$. Call the oils *A, B, C* and *D* and assign one oil to each of the 16 car-driver combinations in the following arrangement:

|       | $Y_1$ | $Y_2$ | $Y_3$ | $Y_4$ |
|-------|-------|-------|-------|-------|
| $X_1$ | A     | B     | C     | D     |
| $X_2$ | B     | A     | D     | C     |
| $X_3$ | C     | D     | B     | A     |
| $X_4$ | D     | C     | A     | B     |

For example, the cell in the row marked $X_2$ and the column marked $Y_3$ contains entry *D*. This means, in the test drive when car $X_2$ is driven by $Y_3$, oil *D* is to be used. The diagram is a plan for 16 test drives, one for each combination of car and driver. It tells us which oil to use in each case.

Examine the arrangement. Each oil appears 4 times, exactly once in each row (for each car) and also exactly once in each column (for each driver). This setup can, in fact, show how each oil performs with each car and with each driver, while still requiring only 16 test-drives, rather than 64.

An arrangement like that in the motor oil example is called a *Latin square experiment,* and the diagram is called a *Latin square*. This particular array is called a 4 × 4 Latin square because it arranges 4 kinds of objects (the oils) in a 4 × 4 square array. Similarly one could consider 3 × 3 or 6 × 6 Latin squares, or any size, provided only that there are the same numbers of rows as columns as labels *A, B,. . . .* The "Latin" property is that each label appears exactly once in each row and in each column. There are several different 4 × 4 Latin squares.

In a Latin square experiment, one first chooses a Latin square of the needed size at random from all the possible squares. Then, in our experiment, the four drivers are assigned at random to the columns, the four cars at random to the rows, and the four motor oils at random to the labels

*A, B, C, D*. After the test drives are completed, the results are handed over to a statistician, who has ways to test whether the results obtained might have been obtained by chance, or show a real difference in the effect of the oils.

In closing, notice that there is a lot of structure underlying a Latin square; not every square array will do the trick. For example,

|       | $Y_1$ | $Y_2$ | $Y_3$ | $Y_4$ |
|-------|-------|-------|-------|-------|
| $X_1$ | A | B | C | D |
| $X_2$ | B | A | D | C |
| $X_3$ | D | B | A | D |
| $X_4$ | C | C | A | B |

looks like a good plan, but three of the drivers each test only three cars, and some of the other combinations are omitted.

# 4 About Standard Deviations

Suppose you wish to estimate the mean of a normal distribution. The best estimator will be the mean of a sample.

However, the sample standard deviation is *not* the best way to estimate the population standard deviation. It is slightly too large. You should divide by $n-1$, not by $n$, when you calculate the variance of the sample.

That is, if the readings are $x_1, x_2, \ldots, x_n$, then the mean is

$$\bar{x} = \frac{\Sigma x_i}{n}$$

(all sums are taken over $i = 1, 2, \ldots, n$).

The sample standard deviation is

$$\sqrt{\frac{\Sigma(x_i - \bar{x})^2}{n}}$$

but the best estimate of the population standard deviation is

$$s = \sqrt{\frac{\Sigma(x_i - \bar{x})^2}{n-1}}$$

The authors of some textbooks apparently think this is too much for you to grasp. They use

$$s = \sqrt{\frac{\Sigma(x_i - \bar{x})^2}{n-1}}$$

as the *definition* of standard deviation. Their definition is wrong. In most practical problems, where the sample size is a hundred or more, this makes no real difference. But there is no reason not to know the real definition.

Very simply: if the problem asks for the standard deviation of a set, use "the $n$ formula,"

$$\sqrt{\frac{\Sigma(x_i - \bar{x})^2}{n}}$$

where, as before,

$$\bar{x} = \frac{\Sigma x_i}{n}$$

is the mean of the set. If you want to find the standard deviation of some larger set, use the "$n-1$" formula

$$s = \sqrt{\frac{\Sigma(x_i - \bar{x})^2}{n-1}}$$

with that same $\bar{x}$.

207

To get a 95% confidence interval for the mean of a normal (or approximately normal) population (such as human attributes like height or weight), take a sample and use

$$\bar{x} \pm \frac{2s}{\sqrt{n}}$$

where

— if the population standard deviation is known, $\sigma$ say, use $s = \sigma$;

— if the population standard deviation is not known, use

$$s = \sqrt{\frac{\Sigma(x_i - \bar{x})^2}{n - 1}}$$

# 5 CONFIDENCE INTERVALS OF POLLS

The following questions and answers help explain why the confidence intervals in published polls may not quite correspond to your calculations.

*The Questions* In middle of 1983, The Gallup Polls asked 1514 adults, "Do you approve of the way Ronald Reagan is handling his job as President?" 41% of the respondents answered "yes."

    **a.** If the poll had used a simple random sample, what would have been the margin of error in a 95% confidence interval?

    **b.** The actual margin of error for a Gallup poll of this same size is 3%. Why does this not agree with your result in part **a**?

*The Answers*

    **a.**

$$\sqrt{\frac{41 \cdot 59}{1514}} = \sqrt{1.59775\ldots} = 1.264\ldots$$

so the margin of error for the 95% confidence level is twice this (two standard deviations), or 2.53%. Newspapers would ptrobably say "two-and-a half percent."

    **b.** Gallup does not use a simple random sample. They employ a multistage stratified sampling design. The margin of error for 95% confidence is somewhat larger for this sampling design than for a simple random sample of the same size.

As a general rule, more complicated sampling methods will give a larger standard deviations than a simple random sample.

# 6 THE ILLINOIS DRIVER'S LICENSE

We tell you how the Illinois driver's license number is calculated.

***The Soundex*** The *soundex* system to encode names was developed in 1918 by Robert C. Russell. It gives a short string to show roughly how a name (or other word) sounds when spoken. To find the soundex of a name:

1. delete any h or w

2. replace letters by numbers:

   0 for A E I O U or Y

   1 for B F P or V

   2 for C G J K Q S X or Z

   3 for D or T

   4 for L

   5 for M or N

   6 for R

3. if two or more adjacent numbers are the same, omit all but one

4. delete the number corresponding to first character of original name if it is still there (that is, unless the letter was h or w)

5. delete any 0's

6. if there are more than three numbers, keep only the first three; if fewer than three, add 0's to the end to make the code up to three numbers

7. put the first letter of the original name on the front.

Examples:

|    | SPRINGER |    | HAMMAN |
|----|----------|----|--------|
| 1. | SPRINGER | 1. | HAMMAN |
| 2. | 21605206 | 2. | 05505  |
| 3. | 21605206 | 3. | 0505   |
| 4. | 1605206  | 4. | 0505   |
| 5. | 16526    | 5. | 55     |
| 6. | 165      | 6. | 550    |
| 7. | S165     | 7. | H550   |

### *The Illinois Driver's License Number*

The Illinois driver's licence number consists of 12 characters.:
The first 4 characters are the soundex of your last name.

The next 3 characters are formed as follows: an arbitrary 3-digit number is assigned for your first name—for example,

140 for Charles, Clara

120 for Carl, Catherine

100 for all other C

and another number (from 1 to 19) is added for your middle initial, if any (1 for A, 6 for F, 18 for T, U or V, . . .)

The last five are formed from your sex and date of birth:
ABCDE means you were born in the year 19AB; the CDE is derived from your birthday. You add 31 for each month before your birthday, and then add on the day of the month. For a woman, increase this by 600.

For an man born on June 12, $(5 \times 31) + 12 = 155 + 12 = 167$.

For a man born on March 1, $(2 \times 31) + 1 = 62 + 1 = 63$.

For a woman born on June 12, $600 + (5 \times 31) + 12 = 767$.

For a woman born on March 1, $600 + (2 \times 31) + 1 = 663$.

**Example** Catherine A. Churchill, born May 3, 1956.

Churchill: C624

Catherine: 120

Middle initial A: add 1, gives 121

Birth year: 55

Birthdate: $4.31 + 2 = 126$

Woman: add 600, gives 726

C624-1215-5726

It is quite possible for two people to get the same number. In that case an *overflow* number is appended. In Illinois these are not printed on the license.

The Illinois state ID (often used as a proof of age or identity by people who don't drive) has a number derived in the same way, except that the letter is put at the end. Ms Churchill would get ID number 6241-2155-726C.

Similar methods are used in several other states. For example, Wisconsin and Florida use the same method, except the date of birth is encoded differently — they both add 40 for each month, and 500 for a woman. For a woman born on June 12, the 3 digits showing her birthday on a Florida license would be $712 = 500 + (5 \times 40) + 12$.

**Illinois Driver's License Tables** If you want to check your own driver's license number, the following tables will be of use.

Here are the standard numbers for some common first names.

| | |
|---|---|
| 020 | Albert, Alice |
| 040 | Ann, Anna, Anne, Annie, Arthur |
| 080 | Bernard, Bette, Bettie, Betty |
| 120 | Carl, Catherine |
| 140 | Charles, Clara |

| | |
|---|---|
| 180 | Dorothy, Donald |
| 220 | Edward, Elizabeth |
| 260 | Florence, Frank |
| 300 | George, Grace |
| 340 | Harold, Harriet |
| 360 | Harry, Hazel |
| 380 | Helen, Henry |
| 440 | James, Jane, Jayne |
| 460 | Jean, John |
| 480 | Joan, Joseph |
| 560 | Margaret, Martin |
| 580 | Marvin, Mary |
| 600 | Melvin, Mildred |
| 680 | Patricia, Paul |
| 740 | Richard, Ruby |
| 760 | Robert, Ruth |
| 820 | Thelma, Thomas |
| 900 | Walter, Wanda |
| 920 | William, Wilma |

Some minor variants may be given numbers from this list — for example, I would guess Will would be classified as 920. However, if your name is not listed, it will usually be given a code corresponding to its first letter:

| | | | | | | | |
|---|---|---|---|---|---|---|---|
| A | 000 | H | 320 | O | 640 | V | 860 |
| B | 060 | I | 400 | P | 660 | W | 880 |
| C | 100 | J | 420 | Q | 700 | X | 940 |
| D | 160 | K | 500 | R | 720 | Y | 960 |
| E | 200 | L | 520 | S | 780 | Z | 980 |
| F | 240 | M | 540 | T | 800 | | |
| G | 280 | N | 620 | U | 840 | | |

For example, Eric should be 200. If you use only the first initial of your first name, this table is used. If your license just says J. Smith you get 420. Even if your name is John Smith, it's 420, not 460 — they go by the name as printed on the license.

For your middle initial, use

| | | | | | | | |
|---|---|---|---|---|---|---|---|
| A | 1 | H | 8 | O | 14 | V | 18 |
| B | 2 | I | 9 | P | 15 | W | 19 |
| C | 3 | J | 10 | Q | 15 | X | 19 |
| D | 4 | K | 11 | R | 16 | Y | 19 |
| E | 5 | L | 12 | S | 17 | Z | 19 |
| F | 6 | M | 13 | T | 18 | | |
| G | 7 | N | 14 | U | 18 | | |

# 7 EVERYONE WINS

To show an extreme case of how the method chosen might affect the outcome of an election in a realistic situation, we shall consider an example of a political party convention at which five different voting schemes are adopted. Assume that there are 110 delegates to this national convention, at which five of the party members, denoted by $A$, $B$, $C$, $D$, and $E$, have been nominated as the party's presidential candidate. Each delegate must rank all five candidates according to his or her choice. Although there are $5! =: 5 \times 4 \times 3 \times 2 \times 1 = 120$ possible rankings, many fewer will appear in practice because electors typically split into blocs with similar rankings. Let's assume that our 110 delegates submit only six different preference schedules, as indicated in the following table (the *preference profile*):

Number of delegates

|               | 36 | 24 | 20 | 18 | 8 | 4 |
|---------------|----|----|----|----|---|---|
| First choice  | A  | B  | C  | D  | E | E |
| Second choice | D  | E  | B  | C  | B | C |
| Third choice  | E  | D  | E  | E  | D | D |
| Fourth choice | C  | C  | D  | B  | C | B |
| Fifth choice  | B  | A  | A  | A  | A | A |

For example, we see from the table that the 36 delegates who most favor nominee $A$ rank $D$ second, $E$ third, $C$ fourth, and $B$ fifth. Although $A$ has the most first-place votes, he is actually ranked last by the other 74 delegates. The 12 electors who most favor nominee $E$ split into two subgroups of 8 and 4 because they differ between $B$ and $C$ on their second and fourth rankings.

We shall assume that our delegates must stick to these preference schedules throughout the following five voting agendas. That is, we will not allow any delegate to switch preference ordering in order to vote in a more strategic manner or because of new campaigning.

We report the results when five popular voting methods are used.

**1. Plurality.** If the party were to elect its candidate by a simple plurality, nominee $A$ would win with 36 first-place votes, in spite of the fact that $A$ was favored by less than one-third of the electorate and was ranked dead last by the other 74 delegates.

**2. Runoff.** On the other hand, if the party decided that a runoff election should be held between the top two contenders ($A$ and $B$), who together received a majority of the first-place votes in the initial plurality ballot, then candidate $B$ outranks $A$ on 74 of the 110 preference schedules and is declared the winner in the runoff.

**3. Hare Method.** Another approach that could be used is holding a sequence of ballots and eliminating at each stage the nominee with the fewest first-place votes. The last to survive this process becomes the winning candidate. We see in our example that $E$, with only 12 first-place votes, is eliminated in the first round. $E$ can then be deleted from our table of preferences, and all 110 delegates will vote again on successive votes. On the second ballot, the 12 delegates who most

215

favored *E* earlier now vote for their second choices, that is, 8 for *B* and 4 for *C;* the number of first-place votes for the 4 remaining nominees is

| A | B | C | D |
|---|---|---|---|
| 36 | 32 | 24 | 18 |

Thus, *D* is eliminated. On the third ballot the 18 first-place votes for *D* are reassigned to *C*, their second choice, giving

| A | B | C |
|---|---|---|
| 36 | 32 | 42 |

Now *B* is eliminated. On the final round, 74 of the 110 delegates favor *C* over *A*, and therefore *C* wins by this method.

**4. Borda count.** Given that they now have the complete preference schedule for each delegate, the party might instead choose to use a straight Borda count to pick the winner. This could be done, for example, by assigning 5 points to each first-place vote, 4 points for each second, 3 points for a third, 2 points for a fourth, and 1 point for a fifth. The scores are:

*A:* $254 = (5)(36) + (4)(0) + (3)(0) + (2)(0) + (1)(24 + 20 + 18 + 8 + 4)$

*B:* $312 = (5)(24) + (4)(20 + 8) + (3)(0) + (2)(18 + 4) + (1)(36)$

*C:* $324 = (5)(20) + (4)(18 + 4) + (3)(0) + (2)(36 + 24 + 8) + (1)(0)$

*D:* $382 = (5)(18) + (4)(36) + (3)(24 + 8 + 4) + (2)(20) + (1)(0)$

*E:* $378 = (5)(8 + 4) + (4)(24) + (3)(36 + 20 + 18) + (2)(0) + (1)(0)$

The highest total score of 382 is achieved by *D*, who then wins. *A* has the lowest score (254) and *B* the second lowest (312).

**5. Condorcet.** In the Condorcet method, each nominee is matched head-to-head with every other. There are 10 such competitions, and each candidate appears in 4 of them. Assuming sincere voting, we can easily see that *E* wins out over:

*A* by a vote of 74 to 36

*B* by a vote of 66 to 44

*C* by a vote of 72 to 38

*D* by a vote of 56 to 54

In this case, the Condorcet method does produce a winner, namely, *E*.

In summary, our political party has employed five different common voting procedures and has come up with five different winning candidates. We see from this illustration that those with the power to select the voting method may well determine the outcome. Moreover, we have not considered the many possibilities for strategic voting or for the formation of political coalitions. These latter possibilities could alter outcomes and greatly complicate the analysis.

# 8 THE GENERALIZED HARE METHOD

In this section we discuss the *quota method,* a generalization of the Hare method for use when electing several candidates are to be elected.

As an example, suppose 24,000 people are to elect 5 representatives from a larger number of candidates. Each person makes a *preferential* vote in which *all* candidates are listed in order. The *quota* is 4,000 votes. (This number is chosen because you can't have 6 candidates each get more than 4,000 first preferences). Every candidate who gets more than 4,000 votes is elected. If a candidate gets more than 4,000 votes, the excess go to voters' second choices, divided proportionally.

For example, suppose $A$ gets 5,000 votes. 2,500 of these have $B$ as second choice, 2,000 have $C$ second, and 500 have $D$ second. Then $B$ gets 50% of $A$'s preferences, because 2,500 is 50% of 5,000. $C$ gets 40%, and $D$ gets 10%.

$A$ has 1,000 preferences to be distributed. $B$ gets 50% of these (500). So 500 more votes are added to $B$'s total. $C$ gets 40% (400), so 400 more votes are added to $C$'s total. And $D$ receives 100 more votes (10%).

Now the vote totals are examined to see if $B$, $C$ or $D$ have exceeded the quota.

If not enough candidates have been elected, and no one has enough votes, the remaining candidate with the *fewest* votes is eliminated. *All* that candidate's preferences are distributed.

This process may result in complicated numbers, fractions, and so on. This is not a problem nowadays, when votes can be entered on computers.

In the general case, say there are $V$ voters and $N$ places to be filled. The quota is

$$\frac{V}{N+1}.$$

At each stage, if a candidate has more first-place votes than the quota, he or she is declared elected and eliminated from the ballots. The successful candidate's surplus votes are distributed among those ballots where he or she took first place. If no one reaches the quota, the candidate with the fewest first-place votes is eliminated.

The generalized Hare method is only used in large elections, but we shall illustrate the process with a small (unrealistically small) example. Say there are 5 candidates for 3 positions. The voter profiles are

| 6 | 6 | 9 | 6 | 3 | 2 |
|---|---|---|---|---|---|
| A | A | C | C | E | E |
| B | B | D | D | C | A |
| E | D | E | E | D | B |
| D | E | A | B | A | C |
| C | C | B | A | B | D |

There are 32 voters, so the quota is 8.

$A$ gets 12 first place votes, $C$ gets 15. So $A$ and $C$ have both exceeded the quota, so both are elected.

217

For example, consider the six ballots shown in the first column of the table. Both *A* and *C* are eliminated from these ballots (and from rest of the election — they are definitely elected). The votes in this column are now *B* first, *E* second, *D* third.

Now *A* has a surplus of 4 over the quota. Since there are six votes in each list, the surplus is divided between them in the ratio 6 to 6. This give 2 votes to each of the lists. For example, the first column is replaced by a list *B, E, D* with 2 votes.

*C* has a surplus of 7. When 7 is divided in the ratio 9:6, the proportions are 4.2 and 2.8. For example, the third column becomes *D, E, B*, and is credited with 4.2 votes. Column 5, on the other hand, retains all its votes. This results in the new preference profile

| 2 | 2 | 4.2 | 2.8 | 3 | 2 |
|---|---|-----|-----|---|---|
| *B* | *B* | *D* | *D* | *E* | *E* |
| *E* | *D* | *E* | *E* | *D* | *B* |
| *D* | *E* | *B* | *B* | *B* | *D* |

*B* has 4 first place votes, *D* has 7, *E* has 5. No one meets the quota. So *B*, having the fewest votes, is eliminated, obtaining

| 2 | 2 | 4.2 | 2.8 | 3 | 2 |
|---|---|-----|-----|---|---|
| *E* | *D* | *D* | *D* | *E* | *E* |
| *D* | *E* | *E* | *E* | *D* | *D* |

*D* now has 9 votes, *E* has 7. So *D* is elected. The overall result is that *A, C* and *D* are elected.

# 9 RADIOACTIVE DECAY

Radioactive decay works like geometric growth in reverse.

Even when a radioactive element is left undisturbed — no other elements are introduced for chemical reactions, no physical changes are imposed — the element will gradually decompose into elements with lower radioactivity. The amount of the original element present will decrease. The rate of decrease is proportional to the amount present, just as in population growth and compound interest.

It is common to express the rate of radioactive decay in terms of the *half-life*. The half-life of an element is the length of time it takes for the amount present to halve. For example, suppose an element has a half-life of 1 year. If you start with 500 grams of the substance, there will be 250 grams one year later, then 125 grams after another year, and so on. After $n$ half-lives, amount $A$ decays to $A/2^n$.

**Example.** An artificial element has a half-life of one hour. You have 450 grams of the element. Approximately how long will it take until only 50 grams is left?

**Answer.** Suppose the answer is $n$ hours. You want to know the value of $n$ so that

$$\frac{450}{2^n} = 50.$$

The easiest way to find this out is to do a little experimental arithmetic:

When $n = 3$, the amount left is

$$\frac{450}{2^3} = \frac{450}{8} = 56.25.$$

When $n = 4$, the amount left is

$$\frac{450}{2^4} = \frac{450}{16} = 28.125.$$

The approximate answer is that the amount present will fall to 50 grams after a little longer than 3 hours.